NATIONAL GEOGRAPHIC KiDS

美
双

Prehistoric Mammals

史前哺乳动物

第三级

懿海文化 编著

马鸣 译

外语教学与研究出版社
FOREIGN LANGUAGE TEACHING AND RESEARCH PRESS
北京 BEIJING

京权图字：01-2021-5130

图书在版编目 (CIP) 数据

史前哺乳动物：英文、汉文 / 懿海文化编著；马鸣译. —— 北京：外语教学与研究出版社，2021.11（2023.8 重印）
（美国国家地理双语阅读. 第三级）
书名原文：Prehistoric Mammals
ISBN 978-7-5213-3147-9

Ⅰ. ①史… Ⅱ. ①懿… ②马… Ⅲ. ①古动物－哺乳动物纲－少儿读物－英、汉 Ⅳ. ①Q915.87

中国版本图书馆 CIP 数据核字 (2021) 第 236730 号

出 版 人　王　芳
策划编辑　许海峰　刘秀玲　姚　璐
责任编辑　姚　璐
责任校对　华　蕾
装帧设计　许　岚
出版发行　外语教学与研究出版社
社　　址　北京市西三环北路 19 号（100089）
网　　址　https://www.fltrp.com
印　　刷　天津海顺印业包装有限公司
开　　本　650×980　1/16
印　　张　37.5
版　　次　2022 年 3 月第 1 版　2023 年 8 月第 4 次印刷
书　　号　ISBN 978-7-5213-3147-9
定　　价　188.00 元（全 15 册）

如有图书采购需求，图书内容或印刷装订等问题，侵权、盗版书籍等线索，请拨打以下电话或关注官方服务号：
客服电话：400 898 7008
官方服务号：微信搜索并关注公众号"外研社官方服务号"
外研社购书网址：https://fltrp.tmall.com

物料号：331470001

Table of Contents

Mammals Long Ago

Earth's first mammals lived long, long ago. Like mammals today, they had hair or fur. Most gave birth to live babies. And they fed the babies milk.

The first mammals were small. They lived among giant dinosaurs. Most mammals came out only at night and hid safely during the day.

Cimolestes
(sim-oh-LESS-tees),
72 to 56 million
years ago

Alphadon
(AL-fa-don), 75 to 66
million years ago

Then, 66 million years ago, all the big dinosaurs died out. A few types of tiny mammals lived on. But with all the big dinosaurs gone, mammals didn't have to hide anymore.

Over millions of years, mammals of many shapes and sizes filled the Earth. Let's meet some prehistoric mammal superstars!

Word Watch

PREHISTORIC: Prehistoric is a time before people wrote things down.

Super Size

Some prehistoric mammals were huge. A large size helps an animal fight off smaller enemies.

Earth's first really big mammal was about the size of today's rhinos. It had long, saber-like teeth. But, it only ate plants!

PLANT-EATER

Uintatherium (you-IN-tah-THEER-ee-um), 50 to 45 million years ago

Andrewsarchus (AN-drew-SAR-kus),
40 million years ago

One of the biggest meat-eating
mammals had a head almost three
feet long. Its jaws were strong
enough to crush bones.

Word Watch

SABER-LIKE: Shaped like a long,
curved sword, or saber

Q When can three hornless rhinos stand under an umbrella and not get wet?

A When it's not raining!

But even larger mammals walked the Earth. The hornless rhino was the biggest mammal that has ever lived on land.

Indricotherium
(in-DREE-co-thee-ree-um),
37 to 23 million years ago

This plant-eating giant was bigger than four elephants put together. It was almost as large as the biggest dinosaurs.

Huge mammals lived in the ocean, too. Some prehistoric whales were as big as the biggest dinosaurs.

One of Earth's first whales was nearly as long as two buses. Its jaws were full of sharp teeth. Fish swam away fast when this huge hunter was nearby.

Basilosaurus
(bah-SILL-oh-SORE-us),
40 to 34 million years ago

Horns Galore

Prehistoric mammals had some of the strangest horns the world has ever seen.

This skeleton of *Arsinoitherium* in a museum shows two huge horns on its snout and two little horns above its eyes.

Arsinoitherium
(are-sih-noy-THEE-ree-um),
36 to 30 million years ago

Q What's the difference between a car and a *Megacerops*?

A A car has only one horn.

Megacerops (meg-ah-SER-ops), 37 to 34 million years ago

Horns could help fight enemies. Special horns could also help animals attract and win mates.

Word Watch

MATE: Either a male or female in a pair

15

Mammal Armor

Some prehistoric mammals had armor to keep them safe. One animal looked like today's armadillo. But it was the size of a small car!

Glyptotherium
(GLIP-toe-THEER-ee-um),
1.6 million to 10,000 years ago

It had a giant dome
of bony armor on its back.

Word Watch

ARMOR: A cover, layer, or shell that protects the body

17

5 FUN FACTS
About Prehistoric Mammals

1

Ceratogaulus (sir-AT-oh-GALL-us), the horned gopher, was the smallest horned mammal that has ever lived. It was a foot long and weighed about 5 to 10 pounds.

2

Entelodon (en-TELL-oh-don) looked like a giant warthog, or pig. It was a fierce meat-eater with jaws as strong as a crocodile's.

Scientists have studied the bodies of woolly mammoths that have been found frozen in the ice in the cold, northern parts of Siberia.

3

4

Elasmotherium (ee-LAZ-mo-THEER-ee-um) was a big rhino with long hair. It had one huge horn on its head.

The giant deer, *Megaloceros* (meg-uh-LAH-sir-us), had the biggest, fanciest antlers of any mammal ever. They were about 12 feet across.

5

Pouches for Babies

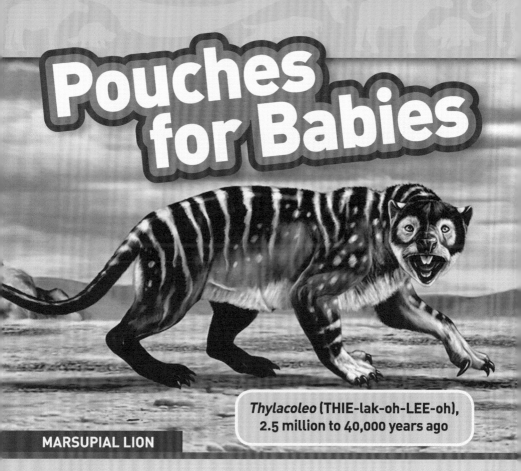

Thylacoleo (THIE-lak-oh-LEE-oh), 2.5 million to 40,000 years ago

MARSUPIAL LION

Many kinds of marsupials (mar-SOO-pee-uls) lived long ago. Marsupial lions were fierce. They had large thumb claws and sharp teeth. They hunted giant kangaroos.

Word Watch

MARSUPIAL: A mammal that keeps its baby in a pouch until the baby grows bigger

Giant kangaroos were much bigger than today's kangaroos. They grew taller than basketball players and twice as heavy. They were fast, too. That helped them escape the marsupial lion's powerful jaws.

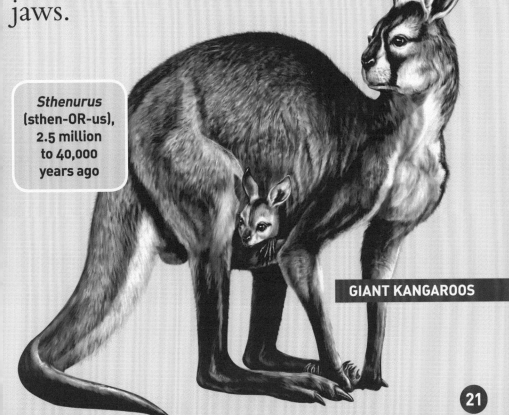

Sthenurus (sthen-OR-us), 2.5 million to 40,000 years ago

GIANT KANGAROOS

Coats for the Cold

About two million years ago, Earth entered an ice age. Winters grew long and very cold.

Many mammals that lived during this time had thick fur coats. These coats helped them stay warm. The woolly mammoths and woolly rhinos were very furry.

Word Watch

ICE AGE: A long period of time, with thick ice sheets covering much of the land

WOOLLY MAMMOTH

Mammuthus (ma-MOO-thuss),
2 million to 10,000 years ago

WOOLLY RHINOS

Coelodonta (SEE-lo-DON-tah),
12,000 to 4,000 years ago

Hunting Skills

The ice age brought many great hunters. The saber-toothed cat used its two swordlike teeth for slicing. These fierce cats hunted horses and bison. They may have tried to hunt giant ground sloths.

Smilodon (SMILE-oh-don), 1 million to 10,000 years ago

Megatherium
(MEG-ah-THEER-ee-um),
4 million to 10,000
years ago

But that would not have been easy!
This sloth was more than 20 feet
tall, with huge clawed hands.

Another great hunter lived during the ice age. Humans! Humans are mammals, too.

Early humans hunted woolly mammoths and woolly rhinos. Humans made weapons of stone. They hunted in groups. Together, they could bring down large animals that fed many people.

Homo sapiens (HO-mo SAY-pee-uns), 200,000 years ago to the present

Over 30,000 years ago, humans painted pictures of large animals on cave walls. Today, most of these giant mammals are gone.

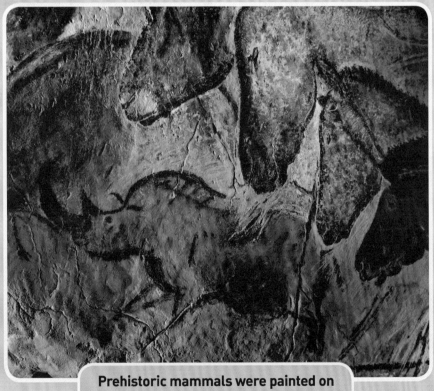

Prehistoric mammals were painted on the walls of the Chauvet Cave in France.

Today's rhinos are relatives of prehistoric woolly rhinos.

As the Earth changed, mammals changed with it. Today, mammals still rule the planet. What animals do you know that are like the prehistoric mammals?

QUIZ WHIZ

How much do you know about prehistoric mammals? Probably a lot! Take this quiz and find out.

Answers are at the bottom of page 31.

1

Which of these is NOT a mammal?

A. A horse
B. A dinosaur
C. A dog
D. A person

2

The largest land mammal that has ever lived was the _____.

A. Elephant
B. Armadillo
C. Hornless rhino
D. Woolly mammoth

Which mammal could be as big as the biggest dinosaurs?

A. A whale
B. A horned gopher
C. A warthog
D. A deer

3

4

A mammal that carries its baby in a pouch is called a _____.

A. Dinosaur
B. Sloth
C. Woolly rhino
D. Marsupial

Which type of animal gives milk to their babies?

A. Birds
B. Dinosaurs
C. Mammals
D. Reptiles

5

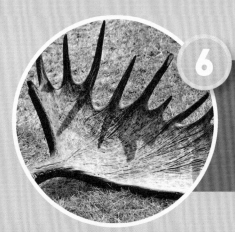

6

Which animal had the largest antlers?

A. Triceratops
B. Woolly rhino
C. Giant deer
D. Giant ground sloth

7

All mammals have _____.

A. Scaly skin
B. Fur or hair
C. Four legs
D. Feathers

Glossary

ARMOR: A cover, layer, or shell that protects the body

ICE AGE: A long period of time, with thick ice sheets covering much of the land

MARSUPIAL: A mammal that keeps its baby in a pouch until the baby grows bigger

MATE: Either a male or female in a pair

PREHISTORIC: Prehistoric is a time before people wrote things down.

SABER-LIKE: Shaped like a long, curved sword, or saber

▶ 第 4—5 页

很久以前的哺乳动物

地球上最早的哺乳动物生活在很久很久以前。和现在的哺乳动物一样,它们长着毛发或毛皮。大多数是胎生的。而且它们给宝宝喂奶。

最早的哺乳动物体形很小。它们生活在巨大的恐龙中间。大多数哺乳动物只在夜间出来活动,在白天则躲得好好的。

白垩窃兽,7,200万至5,600万年以前

▶ 第 6—7 页

后来,在 6,600 万年前,所有的大恐龙都灭绝了。少数几种小型哺乳动物活了下来。但是随着所有大恐龙的消失,哺乳动物不再需要躲躲藏藏了。

几百万年之后,地球上到处都是形状不同、大小不一的哺乳动物。让我们认识一下几位史前哺乳动物"超级巨星"吧!

叉齿兽,7,500万至6,600万年以前

小词典

史前的:史前的是指在人类把事情记录下来之前的时间段。

▶ 第 8—9 页

超大号

一些史前哺乳动物非常大。庞大的身体帮助动物击败比它们小的敌人。

地球上最早的大型哺乳动物和现在的犀牛差不多大。它长着长长的、军刀似的牙齿。但是,它只吃植物!

植食性动物

恐角兽，5,000万
至4,500万年以前

肉食性动物

安氏中兽，
4,000万年以前

最大的一种肉食性哺乳动物有大约 3 英尺（约 0.91 米）长的脑袋。它的颌强劲有力，能把骨头咬碎。

小词典

军刀似的：形状像长长的弯刀或军刀

▶ 第 10—11 页

但是地球上还有更大的哺乳动物。这种无角犀类曾经是陆地上存在过的最大的哺乳动物。

这种植食性巨兽比四头大象加在一起还要大。它几乎和最大的恐龙差不多大。

巨犀，3,700万至
2,300万年以前

▶ 第 12—13 页

海洋里也生活着巨型哺乳动物。一些史前鲸和最大的恐龙差不多大。

地球上最早的一种鲸和两辆公共汽车连在一起差不多长。它的颌上长满了尖利的牙齿。当这个巨型捕食者靠近时，鱼会迅速游走。

龙王鲸，4,000万
至3,400万年以前

▶ 第 14—15 页

好多好多角

有些史前哺乳动物有着世界上曾经出现过的最奇怪的角。

博物馆里的这副重脚兽的骨骼显示，它的口鼻部长着两个大角，眼睛上方长着两个小角。

巨角犀，3,700万至3,400万年以前

重脚兽，3,600万至3,000万年以前

角有助于与敌人战斗。独特的角也可以帮助动物吸引并赢得配偶。

小词典

配偶：一对伴侣中的雄性或雌性

▶ 第 16—17 页

哺乳动物的硬壳

一些史前哺乳动物长着硬壳，以保护自身的安全。有一种动物长得很像现在的犰狳。但是它和一辆小汽车差不多大！
它的背上长着一个巨大的圆盖形骨质硬壳。

小词典

硬壳：保护身体的盖、膜或壳

雕齿兽，160万至1万年以前

关于史前哺乳动物的 5 件趣事

① 圆角鼠，即有角囊地鼠，是曾经存在过的最小的有角类哺乳动物。它长约1英尺（约0.3米），重约5—10磅（约2.27—4.54千克）。

② 完齿兽看上去就像一只巨大的疣猪或猪。它是一种凶猛的肉食性动物，颌与鳄鱼的颌一样强劲有力。

③ 科学家对真猛犸象的尸体进行了研究，它们是在寒冷的西伯利亚北部地区被发现的，当时它们被冻封在冰层里。

④ 板齿犀是一种长着长毛的巨犀。它的头上有一个巨大的角。

⑤ 大角鹿有着所有哺乳动物中最大、最别致的鹿角。它们有大约12英尺（约3.66米）宽。

▶ 第 20—21 页

育儿袋

很久以前生活着很多种有袋动物。袋狮很凶猛。它们巨大的拇指上长着大大的爪子，它们还长着锋利的牙齿。它们猎捕巨型袋鼠。

巨型袋鼠比现在的袋鼠大得多。它们长得比篮球运动员还要高，有他们的两倍重。它们也很快。这有助于它们逃脱袋狮强劲的颌。

袋狮，250万至4万年以前

袋狮

巨型袋鼠

大袋鼠，250万至4万年以前

小词典

有袋动物：在宝宝长大之前，把宝宝养在育儿袋里的哺乳动物

▶ 第 22—23 页

御寒的皮毛

大约 200 万年以前，地球进入冰河时代。冬天变得漫长而寒冷。

生活在这个时期的很多哺乳动物都长着厚厚的皮毛。这些毛为它们保暖。真猛犸象和披毛犀都长着厚厚的皮毛。

真猛犸象

真猛犸象，200万至1万年以前

披毛犀

披毛犀，12,000至4,000年以前

小词典

冰河时代：很长的一段时间，那时厚厚的冰层覆盖着大片的陆地

▶ 第 24—25 页

捕猎技能

冰河时代催生了很多捕猎高手。剑齿虎用两颗剑一般的牙齿撕咬猎物。这种凶猛的猫科动物猎捕马和北美野牛。它们可能试过猎捕大地懒。

但那可不容易！这种树懒超过20英尺（约6.1米）高，长着巨大的爪子。

剑齿虎，100万至1万年以前

大地懒，400万至1万年以前

▶ 第 26—27 页

冰河时代还生活着另一群优秀的猎手。人类！人类也是哺乳动物。

早期的人类猎捕真猛犸象和披毛犀。人类用石头制造武器。他们集体捕猎。他们一起合作可以击败供很多人吃的大型动物。

智人，20万年前至今

▶ 第 28—29 页

3万多年以前，人类将大型动物画在洞穴的墙壁上。现在，这些大型动物大都已经消失了。

法国肖维岩洞的墙壁上画着史前哺乳动物。

现在的犀牛与史前披毛犀是近亲。

随着地球的变化，哺乳动物也在改变。现在，哺乳动物仍然统治着地球。你还知道哪些动物跟史前哺乳动物很像呢？

答题小能手

你知道多少有关史前哺乳动物的知识呢？可能很多！测试一下就知道了。答案在第 31 页下方。

1
下列哪种不是哺乳动物？
A. 马　　　　B. 恐龙
C. 狗　　　　D. 人

2
曾经存在过的最大的陆生哺乳动物是 _____。
A. 大象　　　B. 犰狳
C. 巨犀　　　D. 真猛犸象

3
哪种哺乳动物能跟最大的恐龙一样大？
A. 鲸　　　　B. 有角囊地鼠
C. 疣猪　　　D. 鹿

4
把宝宝养在育儿袋里的哺乳动物叫 _____。
A. 恐龙　　　B. 树懒
C. 披毛犀　　D. 有袋动物

5
哪种动物给宝宝喂奶？
A. 鸟　　　　　B. 恐龙
C. 哺乳动物　　D. 爬行动物

6
哪种动物有最大的鹿角？
A. 三角恐龙　　B. 披毛犀
C. 大角鹿　　　D. 大地懒

7
所有的哺乳动物都有 _____。
A. 鳞状皮肤　　B. 毛皮或毛发
C. 四条腿　　　D. 羽毛

词汇表

硬壳：保护身体的盖、膜或壳

冰河时代：很长的一段时间，那时厚厚的冰层覆盖着大片的陆地

有袋动物：在宝宝长大之前，把宝宝养在育儿袋里的哺乳动物

配偶：一对伴侣中的雄性或雌性

史前的：史前的是指在人类把事情记录下来之前的时间段。

军刀似的：形状像长长的弯刀或军刀